Rubber Industry Advisory Committee

COSHH in the rubber industry
Guidance on the Control of Substances Hazardous to Health Regulations 1988

Contents

Preface *iii*

Introduction *1*

Assessment *1*

The steps to assessment *2*

The prevention or control of exposure to substances hazardous to health *4*

Use of control measures *7*

Maintenance, examination and test of control measures *7*

Monitoring exposure *8*

Health surveillance *9*

Training *11*

Appendix 1: Definition of rubber process dust and rubber fume *12*

Appendix 2: Sample list of processes *13*

Appendix 3: Assessment proforma *14*

Appendix 4: Weekly inspection and maintenance proforma *15*

Appendix 5: LEV record proforma *16*

References *18*

London HMSO

© *Crown copyright 1992*
First published 1992

General enquiries regarding the publications of the
Health and Safety Executive should be addressed to
either of the following HSE Information Centres:

Library and Information Services
Broad Lane
SHEFFIELD S3 7HQ
Telephone: 0742 752539 Telex: 54556

Library and Information Services
Baynards House
1 Chepstow Place
Westbourne Grove
LONDON W2 4TF
Telephone: 071-221 0870 Telex: 25683

ISBN 0 11 885610 3

PREFACE

This guidance has been produced for the Rubber Industry Advisory Committee (RUBIAC) by its Working Party on the Control of Substances Hazardous to Health Regulations 1988 (COSHH), in consultation with the Health and Safety Executive (HSE). The contents have been agreed by RUBIAC which acknowledges with gratitude the work of the Working Party members.

RUBIAC is appointed by the Health and Safety Commission under Section 13(1)(d) of the Health and Safety at Work etc Act 1974 and forms part of the Commission's formal advisory structure. It comprises sixteen members, eight each from the CBI and TUC under HSE chairmanship.

Much has already been written about COSHH. RUBIAC hopes that this guidance, which concentrates on substances and processes used in the rubber industry, will help employers to understand and interpret this important package of measures, so leading to a healthier working environment.

INTRODUCTION

1   The aim of the Control of Substances Hazardous to Health Regulations 1988 (COSHH)[1] is to protect the health of workers who are liable to be exposed to hazardous substances. The employer must ensure that such exposure is either **prevented** or **adequately controlled**. The rubber industry uses more than 500 chemicals in many formulations and at differing temperatures, so the Regulations are of special importance. They cover all workplaces and are constructed on the following logical framework:

ASSESSMENT (Regulation 6); followed by

CONTROL (Regulation 7); which is achieved by

THE USE OF CONTROL MEASURES (Regulation 8); these involve

MAINTENANCE, EXAMINATION AND TESTING (Regulation 9).

EXPOSURE may need to be MONITORED in the WORKPLACE (Regulation 10) and the continuing health of workers checked where appropriate by

HEALTH SURVEILLANCE (Regulation 11).

INFORMATION, INSTRUCTION AND TRAINING (Regulation 12) must be provided at all stages.

2   The Regulations replace the old, piecemeal legislation which dealt with one process at a time. They deal with virtually all substances hazardous to health, and even allow for those where capacity to cause harm has not yet been established.

3   The approach in the Regulations is not new: it brings together the requirements which have been present for many years in health legislation and demands the logical approach, one which a well organised employer would follow in seeking to ensure that employees did not become ill as a result of their work.

4   The Regulations do not simply deal with substances listed as toxic, harmful, irritant or corrosive. They also deal with other hazardous substances, eg sensitisers, certain micro-organisms and dusts of any kind when present in a substantial concentration in air. They require all employers to assess and control the workplace environment in the interests of health.

5   An employer's success in meeting the requirements of COSHH will finally be judged by the control achieved and by the ability to maintain these controls.

6   In the following pages the Rubber Industry Advisory Committee (RUBIAC) advises those who work in the industry on their approach to COSHH. In the context of this booklet the rubber industry is considered to include companies which manufacture flexible foamed polyurethane. The booklet must be read in conjunction with the COSHH General Approved Code of Practice (which contains the Regulations themselves) and the Carcinogens Approved Code of Practice [2]. Much reference will need to be made to these Approved Codes and to other guidance publications from the Health and Safety Executive (see references[3, 7, 10] and [13]).

ASSESSMENT

7   The Regulations hinge on the requirement on the employer to carry out an assessment of the risks arising from work with hazardous substances and of the precautions needed to protect people's health.

8   Regulation 6 states: "An employer shall not carry on any work which is liable to expose any employees to any substance hazardous to health unless he has made a suitable and sufficient assessment of the risks created by that work to the health of those employees and of the steps that need to be taken to meet the requirements of these Regulations." (See the COSHH General Approved Code of Practice[2], paragraphs 11 to 24).

9   This means that **every employer** must evaluate the health risks to employees and others which may result from exposure to any hazardous substances arising from work. In the rubber and polyurethane industry 'substances' will usually mean chemicals, process emissions, by-products and dust and fume, some of which are, or could be, potentially hazardous to health if the appropriate precautions are not taken. Assessment is all about weighing up the size of the health risk from the potentially hazardous chemicals and deciding what to do about them. As well as needing to know about the substances and employee exposure, the assessor needs to know the extent of controls currently in use, and whether they are working efficiently and effectively. This will mean looking at processes on the shop floor in order to identify sources of dust and fume or other hazardous substances, and asking questions about who is exposed, to how much and for how long. If recent occupational exposure data are not available, then air sampling may need to be carried out.

10   If performance data for **existing** engineering controls (eg local exhaust ventilation) are not available, then a thorough appraisal including examination and test will be needed in order to enable the assessor to make the overall assessment.

11 Where the assessment shows the need for additional control measures a 'suitable and sufficient' assessment will identify what is needed and who will carry it out, with a timescale for completion.

12 Where a review of the air sampling data for any particular substance or process shows the need for further personal sampling, then a 'suitable and sufficient' assessment will identify this. The assessment should state the time period for such regular monitoring (see paragraphs 57 to 64).

13 The assessment will also need to include whether there is any need for health surveillance, in what form, how frequently and by whom (see paragraphs 65 to 84).

14 The assessment should extend to consideration of maintenance workers, cleaners and any night shift. The employer's duties extend to other people on the premises such as contractors and to any others likely to be affected by the work. It should be noted that contractors' employers have their own duties under COSHH.

15 In addition, in order to be 'suitable and sufficient' the assessment must identify the action necessary to ensure:

(a) the **use** of control measures;

(b) the appropriate **maintenance**, **examination** and **test** of control measures; and

(c) the provision of **information**, **instruction** and **training** on the COSHH Regulations to employees and, indeed, to anyone carrying out work for the employer.

Who carries out the assessment?

16 From the previous few paragraphs, it will be seen that a 'suitable and sufficient' assessment will be made up of a number of sub-assessments. The obligation to make the assessment falls to the employer, who may be able to do it personally or may delegate **some** of the steps to others in or outside the organisation. The person carrying out the assessment must be competent. Some knowledge of the principles of occupational hygiene is important, if only to identify areas which should be delegated to an outside expert. The booklet *COSHH assessments: a step-by-step guide to assessment and the skills needed for it* [3], considers this question of 'competency'. Managers, operatives and safety representatives will be able to contribute useful information to the person(s) carrying out the tasks because of their knowledge of materials, processes and work practices.

The importance of review

17 Whenever there is a significant change in materials or processes, or new information comes to light, eg on the incidence of ill health, or new developments in control measures, then the assessment will need to be **reviewed**. The assessment will also need to be reviewed in the light of **monitoring** results. It should not therefore be considered to be a 'once and for all' procedure.

THE STEPS TO ASSESSMENT

18 There are a number of practical ways of approaching assessment. In the simplest case - say the factory is making one product at one process, buying in powders and compounding one formulation - then it is a relatively straightforward matter to list the chemicals bought in, add any dusts or fumes that may be formed in the process, add all the cleaning solvents, consider ancillary processes (eg laboratory test chemicals, general cleaning reagents), and begin gathering information about possible hazards, exposure and what control measures may be necessary.

19 More usually the factory will use between 100 to 200 chemicals in a variety of processes (see Appendix 2). Here a convenient approach is to start with a list of the processes, broken down into workstations as necessary, and then list the substances used starting at process number 1. A work sheet or proforma may be useful (see the sample given in Appendix 3).

20 It may be easier to consider some activities separately, for example, laboratory work and perhaps maintenance.

21 The following step-by-step procedure is recommended for each process/workstation, but you may find that the steps have to be repeated in greater detail:

STEP 1

22 List all the substances likely to be encountered.

STEP 2

23 For each substance consult:

(a) any labelling* - does it contain risk and safety phrases and include pictograms of:

   (i) a skull and crossbones? (very toxic or toxic);
   (ii) a St Andrew's Cross? (harmful or irritant);
   (iii) a hand and test tube symbol? (corrosive);

---

* The Classification, Packaging and Labelling of Dangerous Substances (CPL) Regulations 1984[24] and the associated authorised and approved list[25], specify how certain substances should be labelled.

(b) the information provided by the chemical manufacturer/supplier. This will usually be in the form of a safety or hazard data sheet and may well indicate that the risks are greater if the material is heated, milled, or sprayed;

(c) any relevant guidance publications from the Health and Safety Commission (HSC) or Health and Safety Executive (HSE) (for example, *Use of solvents in the rubber industry*[16]);

(d) any relevant trade or trade union literature, such as the British Rubber Manufacturers' Association's Code of Practice *Toxicity and safe handling of rubber chemicals*[19];

(e) standard scientific texts[30, 31] and databases such as HSELINE[32] or RTECS[33];

(f) having consulted the above sources, consider the following:

   (i) does the substance have an LD_{50} of less than 2000 milligrams per kilogram (mg/kg)? (lethal dose (LD) is a measure of acute toxicity);

   (ii) does it have a listed **occupational exposure limit** (OEL)? Guidance Note EH40[7], which is updated annually, gives current limits which refer to inhalation risks. Almost all refer to the average concentrations in air over an 8-hour-day, or over a 10-minute short-term reference period. Some substances, eg rubber process dust (see Appendix 1) and rubber fume, have been assigned **a maximum exposure limit** (MEL), which means that exposure must be reduced so far as is reasonably practicable, and in any case below the MEL. Other substances, eg carbon black, talc, zinc stearate, have been assigned an **occupational exposure standard** (OES), which means that exposure must be reduced to that standard (see paragraph 45(b));

   (iii) is the substance likely to produce higher concentrations of dust in the air than 10 milligrams per cubic metre (mg/m^3) of inhalable dust, or more than 5 mg/m^3 of respirable dust? (both are 8-hour time-weighted averages). You may not be able to answer this question until you carry out Step 5, but bear it in mind until then;

   (iv) even if the substance does not have a listed OEL or the relevant risk and safety labels, does the substance have properties equivalent to those that do?

   (v) is it identified in the Carcinogens ACOP[2] as a substance to which that Code applies? (Note that processes giving rise to rubber process dust and rubber fume are included in the ACOP.)

STEP 3

24 Add to the list any substances generated by the process (eg hot rubber fume, or any other breakdown products arising from the process - which may depend on the processing temperature). Make a note of the conditions which might be more dangerous and look for them during Step 5.

25 Add micro-organisms which create a hazard to health (eg because of contamination of process water cooling, or water cooling towers).

26 Add substances used for cleaning and gather the Step 2 information for these.

STEP 4

27 For those substances which elicit a **yes** answer to **any** of the questions at Step 2 (f) the possible health effects should be noted, with the route by which exposure could occur - ie inhalation, swallowing, skin absorption.

STEP 5

28 Look at the process workstation, the method of work, the sources of dust and fume and existing control measures. Evaluate the degree of exposure in relation to route of entry into the body and record actual exposures where these have been measured, eg by atmospheric sampling or where they are indicated by biological monitoring. Where rubber process dust or rubber fume are present it will usually be necessary to refer to air sampling measurements. Sampling may also be needed for other substances, especially when they have occupational exposure limits (eg some solvents, isocyanates, MbOCA). Is exposure to each substance prevented or adequately controlled? That is, at this workstation, is exposure to substance 'X' prevented or controlled within acceptable limits (dust-suppressed materials help here). Record the assessment of risk at this process from substance 'X', and repeat for all other substances. Also specify and action any need for regular monitoring (see paragraphs 57 to 64).

STEP 6

29 Record what control measures are used at the workstation and whether they obtain adequate control. Typical records might read:

(a) 'fully enclosed and automated plant subject to weekly maintenance schedule as attached';

(b) 'exhaust ventilation achieving face velocity of 0.7 metre per second (m/s) at the booth opening, with duct static pressures as indicated on attached diagram and test log';

(c) 'local exhaust ventilation at three hoods exhausting to dust collection plant fitted with manometer. Training and instruction given to ensure that appropriate action is taken when manometer reading falls to the red line';

(d) 'the following approved respiratory protective equipment is worn for weekly maintenance activities: (eg half-mask respirator (name type) with filters (name type) from supplier (name of supplier), boilersuit and nitrile gloves (state length of gloves); clean shower facilities are also provided'.

30 Record and action any need for additional control measures, and cross reference to the schedule for periodic examination and test of the engineering controls (see paragraphs 48 to 55).

STEP 7

31 Consider whether health surveillance is appropriate. If so, record which employees are subject to it, who is to carry it out and how frequently, eg:

(a) 'chemical 'A' is supplied only as a dusty powder and there is a risk of severe dermatitis despite wearing 'B' type gloves changed daily and rigorous use of good hygiene. The five operatives at this process should therefore be subject to monthly hand and forearm examination by an appropriately trained foreman/nurse or other responsible person and a record kept';

(b) 'process uses toluene di-isocyanate TDI' (see paragraph 73);

(c) 'process uses MbOCA' (see paragraph 75).

32 Repeat all the steps given in paragraphs 22 to 31 for the remaining processes/workstations, making sure that in each case regular cleaning, emergency spillage, maintenance and shift variations are taken into account. Also consider personnel in adjacent departments, contractors and members of the public, where relevant.

33 Assessments will usually need to be written down.

Provision of information

34 The results of the assessment should be explained to the employees or their safety representatives so that they know about the likely risks and precautions needed.

THE PREVENTION OR CONTROL OF EXPOSURE TO SUBSTANCES HAZARDOUS TO HEALTH

35 Regulation 7 of COSHH requires every employer to ensure that the exposure of employees to substances hazardous to health is either **prevented**, or where this is not reasonably practicable, **adequately controlled**. Employers should achieve prevention or control of exposure by measures other than the provision of personal protective equipment, where this is reasonably practicable. 'Exposure' means exposure by inhalation, ingestion, absorption through or contact with the skin.

36 This crucial Regulation deals directly with the prevention or control of exposure to substances hazardous to health. **Assessment**, **training** and **instruction** lead to prevention or control. **Monitoring**, **maintenance** and **health surveillance** show that control has been achieved. Employers' practical success will be judged by the extent to which they create a healthy workplace.

Hierarchy of control measures

37 There is a well-established hierarchy of measures for preventing or controlling exposure, see paragraphs 32 to 36 of the COSHH General ACOP[2]. The employer should adopt this approach which has already been used in other RUBIAC publications and in the British Rubber Manufacturers' Association's Code of Practice[19]. Where a preferred option from the list is not used for any particular process or substance the assessment should make clear why this option was not reasonably practicable. The hierarchy of measures is as follows:

(a) eliminate (for example, by stopping the use of antioxidant Nonox S, and retarder N-nitroso-diphenylamine);

(b) substitute with a less hazardous substance (eg by using toluene instead of benzene), or by the same substance in a less hazardous form (eg ethylene thiourea ETU in a polymer-bound pellet);

(c) totally enclose the process - eg automatic weighing;

(d) partially enclose the process - with local exhaust ventilation;

(e) use local exhaust ventilation;

(f) use general ventilation - eg roof extractors;

(g) use segregation, as in drug weighing or mould cleaning;

(h) clean regularly using a type H industrial vacuum cleaner (type H for dusts hazardous to health[36]);

(i) store and dispose of materials safely;

(j) use safe systems of work - for example, when entering vessels to clean inside, or maintaining internal mixers;

(k) use personal protective equipment - eg for maintenance;

(l) prohibit eating, drinking and smoking in contaminated areas;

(m) provide appropriate washing and changing facilities.

Although washing facilities have been listed last, good facilities and a high standard of personal hygiene are nonetheless very important.

Technical/engineering and operational control measures

38 Although prevention should be explored as fully as possible (by elimination and substitution), it is inevitable that many processes in the rubber industry will require control. Control measures include both **technical/engineering controls** and **operational controls**. Technical/engineering controls should preferably be applied at source, eg local exhaust ventilation, but other engineering control measures, such as powered general ventilation of a workroom may have a place. Operational controls include the way in which the work is organised so as to reduce the hazard, eg by excluding unnecessary personnel from designated areas, introducing work patterns which minimise individual exposures, and operating safe systems of work.

39 Control of dust and fume has been given a high priority by the industry and by RUBIAC for many years. Publications which outline practical measures are available on two-roll mills[14], powder handling and weighing[15], and use of solvents[16]. Some examples of the application of control measures are shown in Table 1 below.

Table 1 Examples of the application of control measures

Process	Control measure
Powder handling and weighing:	
Use of wet products or dust-reduced forms, such as oiled powders, pellets and pre-dispersed forms of masterbatches	Substitution
Choice of the most suitable type and size of packaging, and means of transport	Operational and technical
LEV at sack opening, emptying and disposal	Engineering
Automatic weighing	Technical/engineering
Direct weighing into container (not weigh pan)	Operational
Use of compound compatible bags	Technical
Maintenance of the air cleaning system	Engineering
Emergency spillage procedure	Operational
Internal mixers, dump mills, two-roll mills:	
Local exhaust ventilation	Engineering
Arrangements for maintenance and cleaning	Operational and use of personal protective equipment
Extruders, calenders and vulcanising operations:	
Careful selection of materials	Elimination/substitution
Choice of process temperature	Operational/technical
Application of LEV to freshly cured articles (for trimming and inspection)	Engineering
Putting all scrap inside the exhaust ventilated enclosure, or submerging under water	Operational
Use of LEV at presses	Engineering
Use of LEV at autoclave doors	Engineering
Use of general exhaust ventilation	Engineering

40 The employer and engineers should avoid falling into the trap of adding ventilation control to a poorly-designed process. They should first try to **redesign** the process itself to be as clean as possible, as part of the **control** arrangements.

Personal protective equipment

41 In general terms it is the **process** which should be fitted with the safeguards, **not the operator**. However, protective overalls, aprons, gloves etc can often be a useful and necessary addition to the precautions adopted to prevent direct contact with hazardous chemicals. For example, when working with organic solvents, working methods should be adopted which do not involve direct skin contact with the solvent. Even when such methods have been adopted, impervious aprons and gloves should normally be worn by operatives on these processes, unless it is impractical.

Respiratory protective equipment

42 Respiratory protective equipment (RPE) is usually much less easy to wear than the other items of personal protective equipment and requires a greater degree of attention to ensure that it is fitted correctly and working efficiently. For this reason this type of equipment is certainly to be regarded as a last resort, only to be adopted when all other methods of control have been considered. The main circumstances where RPE is acceptable are:

(a) short-term measures to enable existing operations to continue until more satisfactory control measures are installed;

(b) emergency procedures where urgent action is required owing to failure of the plant;

(c) maintenance procedures where employees cannot be protected in other ways;

(d) short duration operations (less than half-an-hour) which may not justify other types of control.

43 If RPE has to be used for controlling exposure it must be suitable for the purpose and be either type approved by HSE or conform to an HSE approved standard. 'Suitable' means:

(a) the equipment should have the potential to give the required degree of protection;

(b) the equipment should be matched to the wearer and the circumstances of the work;

(c) the wearer should be properly trained to use the equipment;

(d) the wearer should use the equipment properly;

(e) the equipment should be in good condition and be effectively maintained.

44 The required protection will not be realised unless all these things are done. Particular care must be taken where there is a shortage of oxygen or any danger of losing consciousness from fumes etc (as, for example, in confined space work). In such cases breathing apparatus should be worn, **never** filtering respirators. See references[11] and[12] for further information.

Exposure by inhalation

45 'Adequate' control of exposure **by inhalation** will be achieved as follows:

(a) *Substances assigned a maximum exposure limit (MEL), eg rubber fume and rubber process dust -*

   The exposure should be reduced so far as is reasonably practicable and in any case below the MEL, based on the 8-hour long-term or 10-minute short-term reference period. Achievement of this for substances having an 8-hour long-term reference period normally should be demonstrated by a regular programme of monitoring. In addition, short-term (10 minute) exposure limits (where specified) should never be exceeded;

(b) *Substances assigned an occupational exposure standard (OES) -*

   Exposure should be reduced to the OES. Where exposure is above the OES, but this has been identified and steps are being taken to comply as soon as possible, this is acceptable. The time period should be stated in the assessment. Interim use of respiratory protective equipment may be needed;

(c) *Other inhalable substances -*

   Reference should be made to BRMA's Code of Practice[19] or to information supplied by the manufacturer, trade associations, trade unions or HSE. (See paragraph 29 of the COSHH General ACOP[2].)

Exposure by ingestion or by absorption through the skin

46 It is important not to regard compliance with airborne OELs as the only requirement. Solvents are commonly absorbed through the skin and may in that way act as a carrier for transporting other substances into the body. A number of substances listed in Guidance Note EH40[7] have been given a 'skin' notation in addition to an 'inhalation limit'. For some substances such as MbOCA, depending on the conditions of use, skin absorption is the primary exposure route. Reference to suppliers' data sheets and BRMA's Code of Practice[19]

will give information, and appropriate control methods should be applied.

USE OF CONTROL MEASURES

47 Under Regulation 8 **the employer** is required to take all reasonable steps to ensure that the measures taken to meet the Regulations are properly used or applied. **Employees** are required to make full and proper use of control measures, personal protective equipment or anything else provided in order to fulfil the Regulations. Employees are also required to report defects to their employer without delay.

MAINTENANCE, EXAMINATION AND TEST OF CONTROL MEASURES

48 Regulation 9 states: "Every employer who provides any control measure to meet the requirements of Regulation 7 shall ensure that it is maintained in an efficient state, in efficient working order and in good repair." Thorough examination and tests of technical/engineering controls and of certain respiratory protective equipment are required as well as a system for recording the results, and indicating any required repairs. Local exhaust ventilation plant requires examination and test at least once every 14 months. Records must be kept for at least five years.

49 Employers must have comprehensive arrangements in place to maintain all those control measures which existed already or which they have introduced in order to comply with Regulation 7 of COSHH.

50 All engineering control measures in use should be given a weekly visual check and this should be recorded with a note of any action taken. This check should include plant such as vacuum cleaners and fans as well as checks on local exhaust ventilation (LEV). The weekly inspections should also include checking of permanently fitted monitoring devices (see Appendix 4).

51 Preventive servicing procedures should specify which engineering control measures require servicing. The nature and frequency of the servicing should be laid down, who is responsible for organising the servicing, and how any defects disclosed are put right.

52 The engineering control measures (including LEV) should have been given an initial appraisal and test on commissioning to show that the plant was working effectively (ie capturing and disposing of emissions) and was working to its specified performance. There should be reference values (eg air velocities, static pressures). If an initial appraisal was not carried out then it will have to be done before the employer can complete the **assessment**. The employer will not be able to decide if the control measures are effective without this information. Ongoing maintenance is also important,

Regulation 9 'maintenance' requires engineering controls to be thoroughly examined and tested at suitable intervals and the results to be recorded. These results should be compared with the **assessment**, and any defects remedied.

Local exhaust ventilation

53 The statutory periodic thorough examination and test of all LEV plant provided as a control measure is a new requirement for most industries and necessarily will need close scrutiny. Paragraphs 59 to 61 of the COSHH General ACOP[2] give details of the particulars which should be recorded (see also booklet HS(G) 54[10]). Appendix 5 gives an example of some of the details which should be included in a thorough examination and test of LEV plant fitted at, say, powder weighing. Note that there is no prescribed form. A plan of the layout of each system showing the position of hoods and sampling points will be invaluable.

54 The **extent** of the thorough examination depends upon an appraisal made by the competent person at the time of examination. It will require sufficient tests to check that there have been no changes to the plant or process that may affect control, that the plant has not deteriorated and that its performance is unchanged (eg by air pressure and velocity measurement), ie that the person carrying out the examination and test is satisfied that exposure to the substance is being adequately controlled. It depends to some degree on how much data is available from the initial appraisal of the LEV plant. Comparison of, for example, static pressure measurements with the reference value, will give an immediate indication of loss of control if any particular measurement has changed. This is a relatively easy and convenient way of checking performance. Periodic thorough examination of various plant is often carried out by engineer surveyors of insurance companies. Reference to such personnel is not, however, essential and other people may also have the necessary specialist knowledge and professional expertise.

55 LEV systems that return filtered or cleaned air to the workplace pose an added risk because most air cleaners will allow some pollutant to pass through and, should the air cleaner fail, large quantities of pollutant may be released to the workplace. Systems which recirculate filtered air should therefore receive a very high standard of maintenance. The general advice against returning filtered air to the workroom is based on the potentially serious effects if a leak develops in the filters. Employers should critically judge the cost savings in heating against the extra costs of the additional maintenance needed for returned air. The statutory examination and test should also assess the efficiency and integrity of such systems. Techniques for assessing such systems include visual examination, testing the pressure drop across filters, monitoring return air quality

and testing filters for compliance with published standards. Monitoring the return air quality is the most useful but often the most difficult test (especially, for instance, with electrostatic precipitators). It should be noted that the Approved Code of Practice for Control of Lead at Work[26] allows recirculation of lead-contaminated air into a work area only where the concentration of lead in the exhaust air is less than one-tenth of the lead-in-air standard.

Respiratory protective equipment

56 Details of the required examination and test of RPE, and the records to be kept, are given in paragraphs 62 to 65 of the COSHH General ACOP[2].

MONITORING EXPOSURE

57 Sampling for airborne contaminants may be carried out for a number of reasons. Monitoring under the COSHH Regulations has a precise meaning and, where required, should be carried out at least annually. (For some processes, not in the rubber industry, a specific time period is laid down in a schedule.) Employers may wish to carry out air sampling at other times - even though routine monitoring might not be required under COSHH it might be a good idea to carry out air sampling periodically.

58 Regulation 10 states: "In any case in which:

(a) it is requisite for ensuring the maintenance of adequate control of the exposure of employees to substances hazardous to health; or

(b) it is otherwise requisite for protecting the health of employees; the employer shall ensure that the exposure of employees to substances hazardous to health is monitored in accordance with a suitable procedure."

59 The purpose of routine monitoring, as opposed to the initial sampling for the assessment, is to check that the control measures remain effective and are likely to sustain adequate control of exposure.

60 The judgment concerning the need for routine monitoring for any or all of the identified substances hazardous to health should be made during the **assessment**. Each case should be considered on its merits; Guidance Note EH42[8] may assist. The judgment will need to consider factors such as:

(a) the inherent toxicity of the substance;

(b) whether it is potentially cancer causing;

(c) the extent of potential exposure (ie the risk);

(d) when a substance has a MEL, how closely the personal exposure approaches it. If the exposure was, say, less than $1/4$ MEL, and that exposure was as low as is reasonably practicable, then (depending on the other factors) regular monitoring would not be essential. If the results for a particular substance and workstation were all below the MEL with the mean being less than half the MEL, and that level of exposure was as low as is reasonably practicable, then routine monitoring should be considered, perhaps annually. If the mean value was greater than half the MEL, with some results very close to the MEL, and that level of exposure was as low as is reasonably practicable, then routine monitoring at a frequency of more than once per year would be needed; even up to monitoring continuously by a workroom air detector/alarm if the potential harm would be very considerable;

(e) when a substance has an OES, how closely the personal exposures approach the OES. If the results are generally at the OES then routine monitoring (annually) would probably be required.

61 Personal sampling is generally required, but not necessarily of everyone. Where the exposed population is large, one approach is to sample workers randomly (this assumes that all are working in a similar way on the same process). Alternatively, the exposed population may be subdivided into homogeneous groups. Then a number of members of each subgroup (at least one in ten) should be selected for monitoring. Subgroups, which include night shift workers and individual workers who carry out jobs such as cleaning, should not be overlooked. Where available, recommended measurement methods should be used. These are published in HSE's series of publications which cover methods for determining hazardous substances, for example, MDHS 14 *General methods for the gravimetric determination of respirable and total inhalable dust*[27], and MDHS 47 *Rubber fumes in air measured as 'total particulates' and 'cyclohexane soluble material'* [28].

62 Detectors/detector alarms, although not personal exposure monitors, have a significant role to play in ensuring the maintenance of adequate controls, eg leak detectors in ducts and toluene di-isocyanate (TDI) detectors in foam plant.

63 Where the record is representative of the personal exposure of identifiable employees, records must be kept for at least 30 years; otherwise (eg in the case of static sampling or group results) for at least five years. The information should include the date, results, monitoring procedures and duration, along with the names of individuals/locations/job tasks. The records must be able to be readily compared with health surveillance records (see paragraph 80).

64 Records of monitoring must be made available to employees or their safety representatives and, on

HEALTH SURVEILLANCE

65 Regulation 11 begins: "Where it is appropriate for the protection of the health of his employees who are, or are liable to be, exposed to a substance hazardous to health, the employer shall ensure that such employees are under suitable health surveillance." The judgment as to whether there is a need for health surveillance and as to what is suitable should be made during the **assessment**. The term 'health surveillance' covers a number of procedures. The choice of which to adopt depends on the type and extent of risk involved. Some examples are:

(a) medical surveillance by a doctor;

(b) enquiries about symptoms by interview or questionnaire, eg by an occupational health nurse;

(c) periodic inspection by a responsible person (someone appointed by the employer who is competent to carry out the relevant procedure and who is charged with reporting to the employer the conclusions of the procedure);

(d) keeping of records of employees' occupational history;

(e) in the case of workers who require health surveillance - a health record containing particulars approved by HSE, with a copy kept for 30 years from the date of the last entry.

66 The objectives of health surveillance include:

(a) detecting at an early stage adverse health changes which may be attributable to substances hazardous to health;

(b) assisting in the evaluation of the effectiveness of control measures;

(c) collecting data for subsequent evaluation.

67 The requirement under the Indiarubber Regulations 1922 for monthly medical examinations of people carrying out certain processes (ie the 'cold cure' process) has been revoked and carried forward into COSHH (see Schedule 5 of the Regulations, page 30 of the COSHH General ACOP[2]). Hence, where carbon tetrachloride or trichloroethylene are used in the manufacture of Indiarubber, then medical surveillance under the supervision of an HSE employment medical adviser or appointed doctor will still be required. Such medical surveillance will not be needed, however, where exposure is insignificant. 'Indiarubber' is taken here to include all articles or goods made wholly or partially of Indiarubber (ie natural, not synthetic rubber).

68 This situation will be rare. Note that Schedule 5 does not apply to the use of trichloroethylene as a degreaser, or in an adhesive, but health surveillance may still be appropriate in some cases.

69 Health surveillance will be necessary where there is significant exposure to materials which could produce an 'identifiable disease' or 'adverse health effect' and where a 'valid technique' is available for detecting the disease or the effect. There has to be a reasonable likelihood that the disease/effect may occur under the particular conditions of work. Assistance in establishing areas where this would apply may be found by referring to *Health surveillance under COSHH*[13], the BRMA Code of Practice[19] and the COSHH General ACOP[2]. Substances and processes identified in Appendix 1 to the Carcinogens ACOP[2] would be identified here, provided exposure was regarded as being significant. The Carcinogens ACOP applies to all substances which have an R45 risk phrase 'May cause cancer' on the label, as well as to the substances listed by name or process in its Appendix 1.

70 Thus, health surveillance will be appropriate where people are exposed to known or suspect carcinogens, unless that exposure is not significant. Some mineral oils, which may include process oils used for rubber manufacture, and used engine oil are listed in Appendix 1 to the Carcinogens ACOP[2]: regular skin inspection by a suitably qualified person, or regular enquiries by a responsible person about any symptoms following self-inspection by the employees concerned, will be needed unless exposure has been assessed as not significant. (In addition, the workers themselves should be educated about the need to recognise a skin problem and what steps need to be taken to report problems and seek advice. This would fulfil part of the requirement under Regulation 12 to provide information and training - see paragraphs 86 to 92.) Further information on this is given in HSE Guidance Note EH 58 *The carcinogenicity of mineral oils*[29]. With **other known or suspect carcinogens**, the usual health surveillance requirement is the collection and maintenance of health surveillance records.

71 The procedures involved in health surveillance will depend on the substance. For example, workers exposed to substances known to cause **severe dermatitis**, eg resorcinol, IPPD (isopropyl-phenyl-p-phenylenediamine) should have their hands and forearms examined at suitable intervals by a responsible person. Education again is important - see paragraph 70. Enquiries about

symptoms should be made with those exposed to substances (such as isocyanates) known to cause **occupational asthma**, as part of a medical surveillance system. Exposure to substances of recognised **systemic toxicity**, eg cadmium compounds, will require surveillance where there are valid and appropriate clinical or laboratory techniques. This should be under the supervision of a registered medical practitioner. In circumstances where there is no appropriate clinical test available maintenance of a health surveillance record will suffice.

Rubber process dust and rubber fume

72 Health surveillance (in the form of keeping a **health surveillance record**) is appropriate where workers are exposed to rubber process dust and rubber fume (unless that exposure is not significant). RUBIAC considers, however, that even where the employer's assessment indicates that exposure is not significant health surveillance records should be kept for those factories where rubber process dust or rubber fume are present. This recommendation is not a requirement of the Regulations but would be extremely useful in any future health survey. Surveillance should take the form of keeping a health surveillance record for each employee, with details as given in paragraph 76.

Isocyanates

73 Health surveillance of isocyanate workers should include, in addition to the keeping of a health surveillance record, supervision by a registered medical practitioner (usually the company medical officer) and it is recommended that the health surveillance should follow the regime given in *Toxicity and safe handling of di-isocyanates*[21] published by and available from BRMA. This regime recommends routine screening twice a year after initial pre-employment and early employment health assessments. These assessments include routine lung function testing and updating of a respiratory questionnaire.

74 The results of these tests should be included in the employee's works medical record and, along with the health surveillance record, should also be reviewed annually by the medical officer. The lung function testing and completion of the respiratory questionnaire may be carried out by an occupational health nurse.

MbOCA

75 A health surveillance record should be kept for every employee exposed to MbOCA during their work. The use of biological monitoring for assessing urinary MbOCA may be an effective way of identifying individual exposure and can assist the employer in establishing whether further controls are needed. For further information see the *Code of Practice for the Control of MbOCA at work*[22] and also the leaflet *Working with MbOCA*[23]. Results of these procedures should be included in the employee's works medical record and be available for review with the health surveillance record. Employers should, however, exercise caution when interpreting 'negative' data as these may not be a guarantee that 'adequate control' under COSHH has been achieved. HSE is intending to publish further guidance on biological monitoring in the near future.

Health surveillance records

76 The health surveillance record for every employee undergoing health surveillance should contain at least the following particulars: surname; forename(s); sex; date-of-birth; permanent address, including post code; National Insurance number; date of starting present employment; and an historical record of jobs involving exposure to substances requiring health surveillance in this employment.

77 Where additional health surveillance procedures have been carried out, the conclusions, the date on which they were carried out and the name of the person who carried out the procedures, should be entered in the health record. This **should not** include confidential clinical data. It should include conclusions regarding fitness for work.

78 Although not required by the Regulations, RUBIAC strongly recommends that the record also contains the employee's National Health Service number. This would greatly assist epidemiologists in any future health surveys in the industry.

79 The health surveillance record is of course a quite separate document from an employee's normal works medical record card which has details of pre-employment medical and subsequent medical history. The two types of record should be kept independently.

80 The health surveillance record should be kept in a form compatible with, and capable of being linked to, those records required under Regulation 10 for monitoring of exposure (see paragraph 63). For instance, if a worker's (A's) exposure to any substance hazardous to health has been monitored, then that fact must be 'flagged' on his/her health record, so that reference may be made to those results if required. If another person's (B's) exposure has not been monitored, but, for example, he/she does the same job as A, but on an adjacent machine, or works on A's machine but on a shift that was not sampled on any particular occasion and the pattern of work can be considered to be similar, then B's health record should

also be flagged. All these data will be invaluable in any epidemiological study, in order to spot trends. A list of the names of people undergoing health surveillance should be kept.

81   The employer should allow his/her employees access to their own health record. In addition, information on the collective results of health surveillance (but in a form which preserves confidentiality) should be given to employees and their safety representatives.

82   The health surveillance records and occupational history should be reviewed periodically to identify any job-related adverse health effects. This review could be carried out by medical, nursing, safety or personnel staff.

83   The health surveillance record must be kept for at least 30 years from the date of its last entry.

84   An employer who ceases to trade must notify HSE in writing and offer to send the health surveillance records to HSE.

Record format

85   The records required to be kept regarding the **assessment**, **maintenance of control measures**, **monitoring** and **health surveillance**, may be kept in any format, including on computer. Indeed, the storage of data on computer is highly recommended. Many large companies are doing this in-house, and specialised software packages are available. Confidential clinical data, however, must be accessible only to a doctor.

TRAINING

86   Regulation 12 of COSHH places a wide ranging duty on employers to provide training for employees, whatever their position, who come within the scope of the Regulations. The same Regulation also requires the provision of appropriate instruction and information to employees.

87   Regulation 12 specifically states that the employer shall ensure that any person, even if not employed by the company, who carries out work in connection with the employer's duties under these Regulations, such as risk assessment, has the necessary information, instruction and training. Anyone who does any work for the employer in connection with COSHH must therefore be competent. (See reference[3].)

Management training

88   In many cases assessments will be done in-house. In small to medium-sized firms, without a full-time health and safety officer, it will probably be done by supervisors or managers, either from production or personnel, who have had health and safety added to their other responsibilities. Proper training in how to carry out and evaluate an assessment is vital for management staff. Whatever the level of experience, staff carrying out COSHH duties on behalf of management will almost certainly need specific training before they can be considered competent.

89   Training will need to provide the following basic skills:

(a)   a knowledge and understanding of the COSHH 'package' - the Regulations, ACOPs and guidance material;

(b)   a practical knowledge and understanding of their specific duties under the Regulations linked to practical training if they have to carry out risk assessments, air monitoring etc;

(c)   the ability to specify the action that has to be taken to comply with the Regulations;

(d)   the ability to recognise the limits of their own competence and the point(s) at which outside advice and/or services should be sought;

(e)   the ability to communicate effectively, both to management and other employees, their findings and what action needs taking.

Furthermore, the 'competent' person carrying out the COSHH duties such as risk assessment must be given the authority, the time and the facilities to do the work.

Training and instruction for employees

90   Every employee coming within the scope of the COSHH Regulations has the legal right to suitable and sufficient training and instruction. The training must ensure that employees know about the risks to health and can effectively apply and use:

(a)   the methods of control - technical and engineering, and operational;

(b)   the personal protective equipment;

(c)   the emergency measures.

91   Employees should also be informed about any monitoring and health surveillance which is carried out.

92   COSHH does not lay down specific training standards as these will need to be dealt with on an industry or vocational basis.

APPENDIX 1 DEFINITION OF RUBBER PROCESS DUST AND RUBBER FUME (See paragraph 23(f)(ii))

Appendix 6 of EH 40 gives the following definitions.

Rubber process dust is:

"evolved during the manufacture of intermediates or articles from natural, rubber and/or synthetic elastomers. This definition does not include dusts which, for occupational hygiene purposes, can be dealt with individually. In each such case the relevant occupational exposure limit will apply. Otherwise, where a substance with an assigned occupational exposure limit is present in a mixed dust, the exposure limit for that substance will apply, in addition to the rubber process dust limit."

Rubber fume is:

"fume evolved in the mixing, milling and blending of natural rubber or synthetic elastomers, or of natural rubber and synthetic polymers combined with chemicals, and in the processes which convert the resultant blend into finished products or parts thereof, and including any inspection procedures where fume continues to be evolved.

The limit relates to 'cyclohexane soluble material' determined by the method described in MDHS 47 *Rubber fume in air, measured as 'total particulates' and 'cyclohexane soluble material'*."

APPENDIX 2 SAMPLE LIST OF PROCESSES (see paragraph 19).

Factory XYZ Ltd (which makes general rubber goods)

Processes carried out:

Storage

Weighing

Mixing

Milling

Compression moulding

Trimming

Inspection

Packing

Maintenance

Laboratory

Cleaning

Transport and motor vehicle repair

Substances hazardous to health may be used or generated at all these processes. One particular substance may produce 'no' or 'insignificant' risk to health at one machine (because of the particular controls) but 'great' risk at another process.

APPENDIX 3 ASSESSMENT PROFORMA

Example of a completed proforma for summarising the results of an assessment at one work process/workstation

Department ..**Mixing**.. Date of assessment ..**10 January 199-**.. Process ..**Powder weighing**.. Name of person making assessment ..**B. Smith**..

Workstation ..**Number 1**.. General remarks (on cleanliness, tidiness, spill control, leakages, etc) ..**Untidy — spills being walked out**..

..**of department**.. Is workplace exposure monitoring required? Yes/~~No~~ Time period ..**Annually**..

Substances used	OEL mg/m³ or ppm	Health hazard	Route	Recent air sampling personal mg/m³ or ppm
Synthetic rubber	—	None	—	—
Rubber process dust — accelerator vulcanising agent retarder	} 8 mg/m³ MEL	Possible stomach cancer from excess dust	Inhalation and ingestion	Maximum 5.6 mg/m³ (See sampling results of 20 July 199-)
Carbon black	3.5 mg/m³ OES	—	—	Maximum 2.5 mg/m³ (See sampling results of 20 July 199-)
Anti-degradant	10 mg/m³	—	—	Maximum 8 mg/m³ (See sampling results of 20 July 199-)
Pigment	"	—	—	"
Filler	"	—	—	"
Process oil	—	Dermatitis	Skin	—

Control measures used (see LEV maintenance sheet)

Accelerator in granules
Vulcaniser as aqueous dispersion } Dust suppressed
Anti-degradant as Flakes } Forms
Compound - compatible bags
LEV at weighscale
H type vacuum cleaner in department

Are measures satisfactory or are additional steps needed, if so what?

Maintenance sheet A1/1/9- dated 20 January 199-, with work actioned March 199-, indicates LEV operating as designed. Checked with dust lamp December 199-. Satisfactory.
In my opinion adequate control is achieved but regular thorough vacuum cleaning is required along with a weekly tidy and disposal of materials no longer used. Six monthly review needed. Train operators to clean themselves down before leaving department - reminder in general annual refresher training

Signed ..**B. Smith**..

Personal protective equipment used: (if applicable, list - **boilersuit, gloves etc boiler suit, long gloves**.. Respiratory protective equipment used: (if applicable, list) ..**N/A**..

Is health surveillance required? Yes/~~No~~ Give reason ..**Significant exposure to rubber process dust**.. Number of people needing surveillance ..**3**..

Records to be kept by ..**Personnel Department**.. Form of record ..**Computer**..

Training record

Training/information given on:

Operator's name	Health risk	Precautions	Use of control measures	Defect reporting system	Use of PPE	Good hygiene	Air sampling results	Date
Jones	✓	✓	✓	✓	✓	✓	✓	10 Jan 199-
Green	✓	✓	✓	✓	✓	✓	✓	"
Smith	✓	✓	✓	✓	✓	✓	✓	17 Jan 199-

APPENDIX 4 WEEKLY INSPECTION AND MAINTENANCE PROFORMA

Example for local exhaust ventilation

Visual check for week commencing . . . .

Workstation number . . . .

Item	Check for	Satisfactory/ unsatisfactory	Action required	Comments	Signature	Date	Date of remedial work	Signature
All exhaust hoods, booths	Outward signs of malfunction Physical damage Blockages Poor positioning							
Ductwork	Physical damage or wear Blockage							
Dampers	Not closed *(Note: There should be no need to alter damper positions once a system is set up)*							
Air cleaning	Physical damage Regular emptying of collected material Pressure differential *(Note: From permanently installed manometer)*							
Fan	Motor current *(Note: From permanently installed ammeter)*							

APPENDIX 5 LEV RECORD PROFORMA

Example of matters which should be included in the record following a thorough examination and test of local exhaust ventilation (LEV) plant

Name of employer responsible for the plant ....

Address ....

Location of the LEV plant/process ....

Identification of the LEV plant ....

Hazardous substance(s) (*Name and form eg dust, fume, gas, vapour*) ....

Date of last thorough examination and test ....

Are details of the plant specification or initial appraisal available, together with layout plan of the system with test points? (*If not, an appraisal will be required*)

Assessment of:	Detailed comments	Pressure measurements	Air flow measurements	Dust lamp/smoke tube	Repairs required giving time scale
1 Control	- as to whether control is adequate - giving methods used to assess control				- detail as necessary
2 Condition of enclosures and hoods	- as to their condition (and if not in good condition, whether the wear/damage is/will impair effectiveness) - as to whether static pressures and face velocities are within acceptable range	- record static pressures	- record face velocities	- comment, if used	- detail as necessary
3 Ducting	- as to condition (and whether damaged, worn or blocked) - as to whether pressures/volume flow are in acceptable range	- record velocity pressure	- record volume flow		- detail as necessary

4	Dust collector/ filter	- type of filter - as to any damage, leakage, blockage, binding of filter, condition of housing and cleaning mechanism - if a wet filter, whether water levels are correct - as to whether pressure measurements across filter and velocities are acceptable	- record static pressure measurements across the filter	- detail as necessary
5	Fans	- on condition, any build up of deposits or belt slippage - as to whether static pressure measurements are correct - check details of fan against fan/motor specification	- record static pressure	- record air flow	- detail as necessary
6	Recirculation of filtered air	- see 4 above - as to whether air sampling might be needed to assess the extent of dust penetration	- record static pressure	- comment, if used	- detail as necessary
7	General comments	- on the conditions under which the system was tested - comments on any performance not covered above - comments on any part likely to fail before next examination and test - any special considerations			- detail as necessary

8 Exhaust discharge/flue (are arrangements satisfactory, if not why?) ....

9 Date of this examination and test ....

10 Name and job title of person carrying out the examination and test ....
   Signature and date

11 Name and address of the employer of the person carrying out the examination and test ....
   ....

12 Date of next thorough examination and test ....

REFERENCES

This booklet should be read in conjunction with:

1 *The Control of Substances Hazardous to Health Regulations 1988*, SI 1988/1657 HMSO ISBN 0 11 087657 1 *The Control of Substances Hazardous to Health (Amendment) Regulations 1990* SI 1990/2026 HMSO ISBN 0 11 005026 6

2 *The control of substances hazardous to health and control of carcinogenic substances* Control of Substances Hazardous to Health Regulations 1988 Approved Codes of Practice 2nd ed HMSO 1990 L5 (contains the Regulations in full) ISBN 0 11 885593 X

3 *COSHH assessments: a step-by-step guide to assessment and the skills needed for it* HMSO 1988 ISBN 0 11 885470 4

4 *Introducing COSHH* free HSE leaflet 1988 IND(G)65(L)

5 *Hazard and risk explained* free HSE leaflet 1988 IND(G)67(L)

6 *Introducing assessment: a simplified guide for employers* free HSE leaflet 1988 IND(G)64(L)

7 *Occupational exposure limits* HSE Guidance Note EH 40 HMSO 1991 ISBN 0 11 885580 8 (published annually - database also available, see reference 34)

8 *Monitoring strategies for toxic substances* rev ed HSE Guidance Note EH42 HMSO 1989 ISBN 0 11 885412 7

9 *The control of legionellosis including legionnaires' disease* HSE Guidance Booklet HS(G)70 HMSO 1991 ISBN 0 11 885660 X

10 *The maintenance, examination and testing of local exhaust ventilation* HSE Guidance Booklet HS(G)54 HMSO 1990 ISBN 0 11 885438 0

11 *Respiratory protective equipment: a practical guide for users* HSE Guidance Booklet HS(G)53 HMSO 1990 ISBN 0 11 885522 0

12 *Respiratory protective equipment (RPE): legislative requirements and lists of HSE approved standards and type approved equipment* 2nd ed HMSO 1991 ISBN 0 11 885609 X

13 *Health surveillance under COSHH: guidance for employers* HSE Booklet HMSO 1990 ISBN 0 11 885447 X

The following may also prove useful:

14 *Control of dust and fume at two-roll mills* Rubber Industry Advisory Committee booklet HMSO 1986 ISBN 0 11 883850 4

15 *Dust control in powder handling and weighing* Rubber Industry Advisory Committee booklet HMSO 1989 ISBN 0 11 885495 X

16 *Use of solvents in the rubber industry* Rubber Industry Advisory Committee booklet HMSO 1990 ISBN 0 11 885540 9

17 *Substances for use at work: the provision of information* HSE Guidance Booklet HS(G)27 2nd ed HMSO 1988 ISBN 0 11 885458 5

18 *An introduction to local exhaust ventilation* HSE Guidance Booklet HS(G)37 HMSO 1987 ISBN 0 11 883954 3

19 *Toxicity and safe handling of rubber chemicals* British Rubber Manufacturers' Association's Code of Practice 3rd ed 1990 ISBN 0 90 311018 0. Available from: BRMA Health Research Unit, Scala House, Holloway Circus, Birmingham B1 1EQ (Tel 021-643 9269)

20 *Isocyanates: toxic hazards and precautions* HSE Guidance Note EH16 HMSO 1984 ISBN 0 11 883581 5

21 *Toxicity and safe handling of di-isocyanates and ancillary chemicals: a Code of Practice for polyurethane flexible foam manufacture* British Rubber Manufacturers' Association Ltd 1991 ISBN 0 90 311029 6 (For availability see reference 19)

22 *Code of Practice for the control of MbOCA at work*. Available from Hallam Polymer Engineering Limited, Callywhite Lane, Dronfield S18 6XR (Tel 0246 415511)

23 *Working with MbOCA* free HSE leaflet 1989 MS(A)11

24 *The Classification, Packaging and Labelling of Dangerous Substances Regulations 1984* SI 1984/1244 HMSO ISBN 0 11 047244 6

The Classification, Packaging and Labelling of Dangerous Substances (Amendment) Regulations 1986 SI 1986/1922 HMSO ISBN 0 11 067922 9

The Classification, Packaging and Labelling of Dangerous Substances (Amendment) Regulations 1988 SI 1988/766 HMSO ISBN 0 11 086766 1

25 *Information approved for the classification, packaging and labelling of dangerous substances for supply and conveyance by road* (3rd ed) HSC authorised and approved list HMSO 1990 ISBN 0 11 885542 5

26 *Control of lead at work* Approved Code of Practice The Control of Lead at Work Regulations 1980 HMSO 1985 (COP2)(rev) ISBN 0 11 883780 X

27 *General methods for the gravimetric determination of respirable and total inhalable dust* HSE MDHS 14 (rev) HMSO ISBN 0 7176 0343 1

28 *Rubber fume in air measured as 'total particulates' and 'cyclohexane soluble material'* HSE MDHS 47 HMSO 1987 ISBN 0 7176 0250 8

29 *The carcinogenicity of mineral oils* HSE Guidance Note EH58 HMSO 1990 ISBN 0 11 885581 6

30 Patty F A *et al Patty's industrial hygiene and toxicology* 3rd ed J Wiley and Sons (3 volumes) 1978-85

31 Sax N I and Lewis R J *Dangerous properties of industrial materials* 7th ed 1988 Van Nostrand Reinhold (3 volumes) ISBN 0 44 22802 03

32 *HSELINE* A computer database compiled of bibliographic references to health and safety at work publications. Information about access is available via HSE's Library and Information Services, Broad Lane, Sheffield S3 7HQ (Tel 0742 752539)

33 *Registry of toxic effects of chemical substances* (RTECS) is the leading USA source of worldwide toxicity data (National Institute of Occupational Safety and Health (NIOSH) publication). For details contact US Government Printing Office, DHEW Pub. Also available as a database, access from the UK is available through Blaise-link. For details contact: British Library, Boston Spa, Wetherby, W Yorkshire LS23 7BQ (Tel 0937 546600)

34 *Workplace air and biological monitoring database 1991* EH40. This database, available from HMSO, provides details of occupational exposure limits together with measurement methods ISBN 0 11 885607 3

35 *Dust: general principles of protection* HSE Guidance Note EH44 2nd ed 1991 HMSO ISBN 0 11 885595 6

36 BS 5415: Section 2.2: l986 *Specification for vacuum cleaners, wet and/or dry.* Supplement No1 (1986) to BS 5415: Section 2.2: 1986 *Specification for type H industrial vacuum cleaners for dusts hazardous to health*

Availability

HSE free leaflets are available from the Information Centres at: Broad Lane, Sheffield S3 7HQ (Tel: 0742 752539) and Baynards House, 1 Chepstow Place, Westbourne Grove, London W2 4TF (Tel: 071-221 0870)

HMSO publications are available from: HMSO Books, PO Box 276, London SW8 5DT (Tel 071-873 9090)

British Standards are available from: British Standards Institution, Sales Dept, Linford Wood, Milton Keynes MK14 6LE (Tel 0908 221166)

Printed in the United Kingdom for HSE and published by HMSO 40C 1/92